HUMAN LIFE AND SCIENTIFIC EXPERIMENTS

(In defence of the Dignity of Man)

LINUS AKALI I. SDV

Copy right (c) Linus Ihebirinachi Akali

e-mail: linusakali92@yahoo.com or akalilinus@yahoo.com

First Print, 2016

All rights reserved. No part of this book may be reproduced or transmitted in any form or by any means, electronic or mechanical, including photocopying, recording or by any information storage and retrieval system without the written permission from the author.

DEDICATION

This work is dedicated to my late Parents: late Mr and Mrs Alphonsus I. Akali for being the instruments to my existence, and did not consider me as a specie for scientific experiments. I am grateful to them, and may God rest their souls.

Acknowledgements

I am always grateful to count my blessings and to name them one by one. God has been my abilities and capabilities. Without him I am nothing and with him I am an instrument in his hands. I am grateful to God for his sustenance and his faithfulness. I also wish to acknowledge the opportunities given to me by my Religious congregation the society of the divine vocations (the vocationist Fathers and Brothers). Thanks Rev. Fr. Dr Sylvester Nwutu, who taught me in Christian Ethics and really moderated this work. I am grateful to Dr. Justin Ekennia who taught me Biomedical Ethics. Thanks to my Parish Priest Fr Antonio Razzano sdv and the whole community and parishioners of the SS Annunziata and Maria A Toro in Cava Dei Tirreni. I am grateful to Sr Trinitas Ejikeme sdv, for painstakingly reading and writing the Foreword to this book. I am grateful to my biological family for staying together in this time of great trial of the lost of our dearly beloved mother. I pray God almighty to strengthen you all. I also wish to show my indebtedness to all who have helped me so far in my publications and those who find solace in reading my works, I am grateful to you all. Thanks to my dearly beloved friends, and those whose names I can't mention here, you are all recognized and remembered. I wish you all well in Christ Jesus.

Fr Linus Akali I. sdv

TABLE OF CONTENTS

TITLE

PAGE..1

DEDICATION...3

ACKNOWLEDGEMENTS..4

TABLE OF CONTENTS...5

FOREWORD..7

INTRODUCTION..9

CHAPTER ONE

1.0 JOHN PAUL II'S CONCEPT OF MAN IN EVANGELIUM VITAE..12

    1.1 Man As The "Imago Dei"..12

    1.2 The Sacredness And Inviolability Of Human Life..14

    1.3 The Right To Life Of Every Human Being...17

    1.4 The Duty To Protect The Human Life..19

CHAPTER TWO

2.0 SCIENCE AND THE ARTIFICIAL REPRODUCTIVE TECHNOLOGY..23

    2.1 IN VITRO FERTILIZATION AND EMBRYONIC TRANSFER..24

        2.1.1 Homologous In Vitro Fertilization..26

        2.1.2 Heterologous In Vitro Fertilization..26

    2.2 ARTIFICIAL INSEMINATION..27

        2.2.1 Homologous Artificial Insemination..28

        2.2.2 Heterologous Artificial Insemination..29

    2.3 SEX SELETION..29

    2.4 GAMETE INTRAFALLOPIAN TUBE TRANSFER..31

CHAPTER THREE

3.0 CLONING OR AGAMETIC REPRODUCTION..34

   3.1 Overview Of Cloning..34

   3.2 Animal Cloning..38

      3.2.1 THE BENEFITS OF ANIMAL CLONING..41

   3.3 HUMAN CLONING..43

      3.3.1 Human Therapeutic Cloning..44

      3.3.2 Human Reproductive Cloning..47

      3.3.3 Embryonic Stem Cell Harvesting..49

      3.3.4 Cloning And Surrogate Motherhood..51

   3.4 SCIENTIFIC AND MEDICAL ASPECTS OF HUMAN CLONING..52

CHAPTER FOUR

4.0 ETHICAL EVALUATION OF HUMAN CLONING..56

    4.1 The Catholic Church's Perspective..56

    4.2 Cloning And Human Sexuality..61

    4.3 Human Procreation And Marital Acts..64

    4.4 The Dignity Of Christian Marriage..67

    4.5 The Dignity Of Human Embryo..71

CHAPER FIVE

5.0 CALL TO LIFE AND THE OBLIGATION TO DEFEND LIFE..75

Foreword

Just because we are breathing doesn't necessarily imply that we are fully alive. Life is an exciting and challenging affair, and comes with the freedom and power of decision making. God created us and allowed us to be free but the freedom accorded us by God is a responsible freedom. In our exercise of our God-given freedom we must be mindful of the responsibility and accountability that comes with it which is not diminished by ignorance and fear .The right to exercise freedom is one of the fundamental human rights that must be accorded to everyone and respected, because it gives value to our human existence and dignity. As you read along, the author succinctly, analyzed and simplified how we can be helped to exercise true freedom through the creation story. Human Dignity it is argued can be categorized in three:

(a) Form of Positive or Directive... "treat others as one would like others to treat oneself".
(b) Negative or Prohibitive... "In ways one would not like to be treated, don't treat others that way."
(c) Empathic or Responsive...."Wish upon others what you wish for yourself..."

The golden rule or law of reciprocity is the only way to understand human personality and without it human rights are meaningless and even ceases to exist. The fundamental value of solidarity linked to human dignity focuses the attention on people who are unable to provide for themselves and to the need to live. Worthy of consideration is the argument, that those who have the means, created to provide help and support, either voluntarily or contractually committed, must ensure that forms of solidarity are not exploited or subjected to abuse. Justice is one of the concrete ways in which respect for human dignity finds its expression and equality and the law is a direct consequence of the right to respect for human dignity. Human dignity and human rights are intimately and indissolubly interwoven because human dignity is the foundation of human rights which implies that human rights exist as a consequence of human dignity. The book will serve as an excellent introduction to the ideas of genetic engineering, euthanasia and the ever nagging issue of abortion, therefore I exhort you all to read the book thoughtfully and enjoy it. Though this may not be the last word on the fundamental problem of human right but it a bold and right step in the right direction towards the establishment of the basic principles of mutual respect.

Rev. Sr Trinitas Ejikeme
Congregazione Religiosa
Suore Delle Divine Napoli
Pianura Napoli
Italy

INTRODUCTION

Not really going too deep in the etymology of man, I will rather give a little view on why a human being deserve to be treated differently from all other beings. There are many maxims on the human being and the study of man. Philosophers, theologians, physicists, biologists, psychologists, scientists, economists and other spheres of human endeavours have their various definitions but in all, the lowest common factor is that the human being is the dominant master of all the created or existing things in the whole world. Of all other creatures he is the one that has control over the world. He is the beauty of life and the motivator of life. The human person is an art of God imprinting himself among the created things, because man is made in his image and likeness.

For us Christians, he has the control over all things, being the power and authority given to him by the creator of the universe himself: "And God blessed them. And God said to them, *"Be fruitful and multiply and fill the earth and subdue it, and have dominion over the fish of the sea and over the birds of the heavens and over every living thing that moves on the earth."* (Gen 1:28). Thus man has the power to dominate all other created things. He is also made in the resemblance of the creator, being created in his image and likeness. God's injunction of being fruitful and multiplying in order to fill the earth is the provision for man's participation in the art of creation.

Thus, we can say that man is so import to God in the art of creation. Procreation is the part man has to play in order maintain the work God has begun which has no ending. *"God having just finished all His creation, ending with His masterpieces, the very first man and woman, when He told them to be fruitful and multiply (Genesis 1:28). The world was now fully created with days and nights, seasons and years, plants and animals, and Adam and Eve; and God set in motion His plan to fill the world He created with people (Isaiah 45:18). The world was Adam and Eve's inheritance to fill, and, as stated in the beginning of Genesis 1:28, it was God's blessing for Adam and Eve to have children and work the earth. Commentator Matthew Henry wrote that God blessed the first couple with "a numerous lasting family, to enjoy this inheritance . . . in virtue of which their posterity should extend to the utmost corners of the earth and continue to the utmost period of time."*

I stumbled into a blog doing the course of my research that as that *God is present in the creation of every human life*. The most vivid depiction of this is seen in Psalm 139:13-18. Here the psalmist David expressed his gratitude of being created. The fact that God sovereignty superintended David's creation caused him to praise God. David also pointed out the fact that God had the details of his life planned before eternity. In Jeremiah 29:11 God confirms David's thoughts: " *'For I know the plans I have for you,' declares the Lord, 'plans to prosper you and not to harm you, plans to give you hope and a future.'"* Of course, this brings up a very good question. What about those conceived out of rape or illegitimacy? The parent or parents that are responsible for that child may not "feel" as though that child is a blessing from God.

Using the human person as a spacemen for scientific experiments gives a question mark on the thoughts about man and his dignity. If God has plans for every human being, even before he was conceived, it then springs up the argument of why children are killed unborn in order to save an already existing life or in order to scientifically produce other children? This is an aged-long problem in the sphere of scientific growth in the world, but it is a topic that affects the human person no matter in what we believe in. It does not affect Christians alone nor does it negate the atheists. The respect to life is an obligation to all living being, though the Christians for ages have championed the cause through prolife activities. This work is a call to put heads together and fight in defence of the dignity of man. Lines should be drawn on the use of other creatures in scientific experiments and human beings. This work also argues on the process of conception of the human being against the camp that believes the process of conception does not matter. It is true that God has a plan for every human person but man should not take laws into their hands, trying to defy the divine natural law in order to achieve their quest. This work is geared to provoke many response but it is a call for objective defence of life and an obligation to obey the injunction of procreation. Let us not forget that our God is the lover of life.

"To all the members of the Church, the people of life and for life, I make this most urgent appeal, that together we may offer this world of ours new signs of hope, and work to ensure that justice and solidarity will increase and that a new culture of human life will be affirmed, for the building of an authentic civilization of truth and love."

John Paul II, Evangelim Vitae, para. 6

CHAPTER ONE

1.0 JOHN PAUL II'S CONCEPT OF MAN IN ENVANGELIUM VITAE

1.1 MAN AS THE "IMAGO DEI"

In his encyclical Evangeluim Vitae, Pope John Paul II, establishing the scriptural foundations of the origin of man in Gen 1:27, emphasizes the nature of man. The Old Testament writings form the basis of the Encyclical which the Pope calls the gospel of Life. The Scripture says:

> "God said,' Let us make man in our own image and likeness. Let them rule over the fish of the sea, over the birds of the air, over the cattle, over the wild animals, and over all creeping things that crawl along the ground. So God created man in the image of himself, in the image of God he created him, male and female he created them'."[1]

---
[1] Gen. 1:26-27

Human beings then take the life of God whose image he is. The initiation to create man in the image of God (Gen1:26) is for the special purpose of becoming God's representative, taking charge of all created things and continuing in the creation of the world (continuous creation), especially procreation of life. (Gen 1:28). This shows how much God has valued and loved man as his creature. To this the Pope says: "... to show reverence and love for it, it is a task which God entrusts to every man, calling him as his living image to share in his own Lordship over the world"[2].

This shows that God did not just create man as any other animal or creature but He has a great interest and purpose for creating him. Man is the image of the unseen God. Man, not only becomes the image of God, but shares in God's Lorship"[3]. The Pope calls this a "special participation of man and woman in the creative work "of God"[4]. This is the depth and breadth of the Lordship which God bestows on man. The Pope makes it clear that man is not living the life which is his own but a life given to him as a gift from God. He is not the image of himself but the image of God which is transmitted to him in

---

[2] John Paul II, 1995, "Encyclical: The Gospel of Life (Evangelium Vitae)" No. 42
[3] Ibid, No. 43
[4] Ibid

procreation. Thus the Pope affirms: "In Procreation therefore, through the communication of life from parents to child, God's own image and likeness is transmitted, thanks to the creation of the immortal soul."[5]

Thus the image of God is not only seen in the first man and woman (Adam and Eve) that God created. This image is transmitted in human generations. Insofar as man has life, he is the image of God and in begetting, this image is transmitted. The Pope sees "Begetting" as the continuation of creation.[6] Furthermore, the Pope emphasizes that man, as the living image of God, is willed by his creator to be ruler and Lord. Quoting Saint Gregory of Nyssa the Pope writes:

> *"God made man capable of carrying out his role as King of the earth... Man was created in the image of the one who governs the universe. Everything demonstrates that from the beginning man's nature was marked by royalty... Man is a king."*[7]

---
[5] Ibid
[6] Ibid

In this sense, man acts as he whose image he is. God is Lord, then man is made Lord. God is King and man is made king. He was made to exercise dominion and authority, on God's behalf, over all created things. He is ruler and Lord not only over things but especially over himself, and in a certain sense "Over the life which he has received and which he is able to transmit through procreation"[7].

But the Pope asserts that man must exercise wisdom and love in this duty of dominance. He is to share in the boundless wisdom and love of God which comes, according to the Psalmist, through obedience to God's holy Law" a free and joyful obedience. (Ps 119).

1.2 THE SACREDNESS AND INVIOLABILITY OF HUMAN LIFE

For the fact that man is the image of God and shares in the kingship and Lordship of God makes the human Life sacred. Man is the divine archetype in the sense that life in man is the breath of God. Human nature is intrinsically a thing of the divine. Sacredness here implies that the Human life is untouchable because it is the life of God which is pure and incorruptible.

---

[7] Ibid, No. 52

The Pope writes:

> *"Human life is sacred because from its beginning, it involves 'the creative action of God', and it remains forever in a special relationship with the creator, who is its sole end. God alone is the Lord of life from its beginning until its end; no one can, in any circumstance, claim for himself the right to destroy directly an innocent human being."*[8]

This is to say that it is God alone who has the duty to give or take away human life. Man has no power over the existence of a human life. This means that the human life is equally inviolable. Human life is given the sacred and inviolable character, which reflects the inviolability of the creator himself. The scripture backs the inviolability of the human life with the injunction, "You shall not kill" (Cf) Ex 20:13; Dt 5: 17). In other words, anyone who destroys life, no matter the reason is of the devil. The Pope cites the Gospel of John in this that "He who is a murderer from the beginning, is also a liar and father of lies'. (Jn 8:44)

---

[8] Ibid

Pope John Paul II also emphasized on the love of human as a means of appreciating this gift of life. The love of life is not restricted to the love of the self but to the love of all that is human, whether healthy or sick, old or new, born or unborn, He says:

> *"To kill a human being, in whom the image of God is present, is a particularly serious sin. Only God is the master of life. Yet from the beginning, faced with the many and often tragic cases which occur in the life of individuals and society, Christian reflection has sought a fuller and deeper understanding of what God's commandment prohibit and prescribes."[9]*

What the Pope is saying here is that God, who is the author and giver of life, is the master of life. Thus it is left for him to do with life as he wills. But Christians seem not to fully understand the intricacies of God's laws. They seem not to know, in its fullness, what is forbidden and what is acceptable in God's injunctions. Though the Pope understands that there

---

[9] Ibid

are situations in which what is valuable as proposed by the Laws of God seem to involve Paradox[10], yet he emphasized that the intrinsic value of life and that obligation to love the "self" less than others is based on a true right to self-defense[11] and value when it refers to the innocent person. Thus the "absolute inviolability of innocent human life is a moral truth clearly taught by the sacred scripture"..."[12] It is not only the sacred scripture, the tradition of the church and her magisterium uphold the sacredness and inviolability of human life. Voluntary killing of an innocent human being is by all means held as grave immorality. The Pope cites the tradition when he says:

> *"Nothing and no one can in any way permit the killing of an innocent human being, whether a foetus or an embryos, an infant or an adult, an old person, or one suffering incurable disease, or a person who is dying".*[13]

This means that no matter the state of any human being, since he/she is still alive, he/she is untouchable. No one has any right to terminate anyone's life whether with the person's

---
[10] Ibid No. 55
[11] Ibid
[12] Ibid
[13] Ibid No. 34

consent or not Pope John Paul also extended this to the individual who may wish to temper with his or her own life. He says it is not permitted to seek for destruction of his/her own life or the life of another who was entrusted to his/her care. For the fact that life is of God makes it sacred; and the fact that life is sacred makes it untouchable. No one should then destroy life and no one should ask for or permit the termination of any human life. This is because man has a sublime dignity, a dignity that bonds him to his creator, and that which makes him shine forth as a reflection of God himself.[14]

1.3 THE RIGHT TO LIFE OF EVERY HUMAN BEING

The right to life of every human being has generated much controversy as regards when the human being is due to be accorded the right of living as a human person. This goes back to the question of when the human life starts. But the Pope lets us understand that once God has created the human being, his life as a human being must be respected.

Each person or human being has the right, either to remain as a human being or a human person. He cites the scriptural passage which says that "From man in regard to his

---

[14] Ibid

fellow man I will demand an accounting for human life" (Gen 9:5). Since God is the giver of life, and since life is a free gift to every human being, all have right to life, and no one should deprive another of this divine gift. Though the scripture also sees man as dust, it also clearly states that God cares for him. (cf Ps 8) among the fundamental rights of the human person is the right to life; which takes effect from the moment of conception until death.

The Pope urges that this life should be respected. He regrets the "new cultural climate hostile to life" where many people justify certain crimes against life in the name of liberty; and in many nations around the world laws have been adopted that legalize such practices. These practices are seen in abortion, euthanasia, and other forms of artificial reproduction. Though artificial reproduction (as we are going to discuss later) seems at the service of life, yet it open the door to new threats to life.

Disposing of "spare" embryos or using them for research does not in anyway accord the embryo right to life. Prenatal diagnosis too often is an occasion for procuring an abortion. The newborn babies with serious handicaps or illnesses are sometimes denied nourishment or treatment. Thus Pope John

Paul II says that it is disturbing that crimes like these should be regarded as expressions of individual freedom and rights. He says that concern for real human rights reflects moral sensitivity, but offenses against the lives of the weak and vulnerable human beings threaten human rights.

Also John Paul II states that the problem is rooted in the view that rights where only in those who are competent and autonomous; the weak and dependent, for instance, the unborn and the dying, are vulnerable. Also pointing to freedom another cause of this disrespect to human life, the Holy Father says is the:

> *"Notion of freedom which exalts the individual in an absolute way, and gives no place to solidarity, to openness to others and service of them... a misunderstanding of freedom which leads to a serious distortion of life in society.*[15]

This shows that claims to freedom of such people who are involved in practices that undermine the existence of life are evil. One cannot claim freedom to deprive another of

---
[15] Ibid, No 34

his/hers. No one has right to destroy other people's rights. Life is not such a value to be tempered with. To be alive is a mark of the creator's benevolence. Thus, the Holy Father rejects the move to artificially reproduce life for the gratification of selfish interest, and argues that no one should claim the right or freedom to artificially create life. The dignity of the human person should be respected and as such, the Pope says:

> *"The theory of human rights is based precisely on the affirmation that the human person, unlike animals and things, cannot be subjected to domination by others.... When freedom is made absolute in an individualistic way, it is emptied of its original content, and its very meaning and dignity are contradicted."*[16]

This is to say that clear distinctiveness should be placed in matters concerning the human person as against animals and things. It is only the human person who has the right and dignity of life, and this should be given to him.

---
[16] Ibid

1.4 DUTY TO PROTECT THE HUMAN LIFE

Given the fact that human life is a gift from God, sacred and inviolable, it is then important that every human life be adequately protected. The Human person, whether as an individual, group or society, should have the responsibility of protecting life. The dignity and integrity due to the human person should be upheld. By no means should life become a "means" to an end since life is an end itself. The issue of life and its defence and promotion is not a concern of Christians alone, every human being has the duty to protect life bearing in mind that life is precious and should be held tenaciously. In the document also the Holy Father emphasizes:

> *"Every crime against life is an attack on peace, especially if it strikes at the moral conduct of people... But where human rights are truly professed and publicly recognized and defended, peace becomes the joyful and operative climate of life in society."*[17]

---
[17] Ibid, No 19 - 20

This is to say that the respect and defence of every single life in the society brings peace and harmony to the society. If everyone's life is cared for, then each person sees the reason to belong to the society. But the problem is not the assertion of values such as dignity and integrity of the person, justice and peace, it is building such solid foundations that will guarantee these assertions. The Pope emphasizes that the state should help the Church in the promotion of this primary duty – protection of human life.[19]. Thomas J. O'Donnell says that such attack on life is an "ultimate violation of the divine prerogative of the Author of life"[20].

Also asserting the view of the Roman Pontiff on this Issue, Justine N. Ekennia says:

> *"Every human being has the natural instinct of self-preservation. We all like to preserve our own lives and run away from anything that can harm or terminate our lives".*[18]

This means that each individual naturally tends to protect his/her life from any danger. Yet there are people who commit suicide without thinking of what will become of their lives. The Holy Father insists that everybody has the duty to protect both

---

[18] Ibid, No 137

one's life and those of others. He says that "legitimate defense can be not only a right but a grave duty for someone responsible for another's life".[19]

But the defence of life is not only against taking away life, also it is important to note that those who reproduce life artificially are doing more harm to life than good. Thus Pope John Paul II calls on every family, as "the domestic church" to appreciate children as the gift of God while defending the life of their children- both born and unborn. For him, children are the free gift of their (couples) mutual love,[20] and this "Love, as a sincere gift of self, is what gives the life and freedom of the person their truest meaning.[21], Life is beauty of the human person and as such, everyone should hold on to this beauty because when this life is endangered, the beauty of the person has been defiled.

---

[19] Ibid, No 136
[20] Thomas J., O'Donnell SJ, "Medicine and Christian Morality (3rd edition) the society of S. Paul
   Alba-House, New York, 1996, p. 54.
[21] Justin N. Ekennia, "Bio-Medical Ethics: Issues, Trends and Problems, Bartoz Publishers Inc.,
   Owerri, Nigeria, 2003, p. 95.

In any event, methods that fail to respect the dignity and value of the person must always be avoided. I am thinking in particular of attempts at human cloning with a view to obtaining organs for transplants: these techniques, insofar as they involve the manipulation and destruction of human embryos, are not morally acceptable, even when their proposed goal is good in itself. Science itself points to other forms of *therapeutic intervention* which would not involve cloning or the use of embryonic cells, but rather would make use of stem cells taken from adults. This is the direction that research must follow if it wishes to respect the dignity of each and every human being, even at the embryonic stage.

Pope John Paul II, Address to the 18th International Congress of the Transplantation Society. . . (2000), no. 8

CHAPTER TWO

1.0 SCIENCE AND THE ARTIFICIAL REPRODUCTIVE TECHNOLOGY

As we have already stated in the previous chapter, man is created in the image and likeness of God (Gen 1:27). He is made to be in-charge of the whole created order, including himself. The dominion is to use created things for his own benefit and well being. But man owes the account of his life to God who is the sole owner and giver of life. To aid life, God gave him intellect to manipulate nature for the sustenance of his life. But today, we can see that man has gone beyond manipulation of nature to using his own life as specimen for experiments. Man no longer simply dominates his environment but wants to be lord of his life, he wants to determine his life.

Human beings, through science and technology now produce human life and eliminate it at will. This has to do with all the present-day uses of technologies in human and animal reproductions – artificial reproduction or otherwise, called, "Assisted Reproduction Technologies". These are technologies used to facilitate reproduction in animals as well as human beings, and are also used as therapy, to treat infertility cases. Such technologies include the following. In vitro fertilization, artificial insemination, intrafallopian tube transfer and, of course, human cloning. The question now is the intrinsic meaning of life and the morality of these practices. But before talking about the intrinsic nature of human life as regards artificial reproduction, let us first briefly address the specifics of these technologies.

1.1   IN VITRO FERTILIZATION AND EMBROYONIC TRANSFER

This is one of the technologies whereby human life is manufactured outside the natural process of sexual intercourse between couples or male and female mammals as the case may be.

> *"In Vitro fertilization refers to the technique for conception of a human embryo outside the mother's body; and*

the resulting embryo is implanted in the uterus for gestation"[22]

In Vitro fertilization here implies the opposite of in vivo (conception inside the Female womb). The resultant baby could otherwise be called "test tube baby"[23] it is a process by which egg cells are fertilized by sperm outside the womb in a test tube or Petri dish. It involves hormonally controlling the ovulatory process, egg is fertilized, the resultant zygote is then transferred to the patients' uterus. This is done with the intention of having a successful pregnancy. In Embryonic transfer, the woman is faced with the choice of number of embryos to be planted in her womb[24]. This allows a womb to carry 3 to 6 embryos at the same time. The emphasis here is the fact that fertilization is not natural as when the couple unite sexually through coitus after which fertilization occurs in the woman's womb (in vivo fertilization). This is why babies born through in vitro fertilization are called test tube babies, a colloquial term which refers to the tube –shape containers of glass or plastic resin called "test tube". These kind of instruments are commonly used in chemistry cabs and biology labs. It also explains the etymological meaning of the term "In

---

[22] Justin N. Ekennia, Opt Cit, p. 121-122
[23] http://www.arhp.orgleloning
[24] Report of Advanced fertility centre of Chicago, in vitro fertilization, IVF and infertility treatment in 2001 in http://www.advancedfertility.com

Vitro" which is Latin translation of "within the glass"[25]. The other instrument used in vitro fertilization is called "Petri dish".

Explaining more clearly the process and intention of in vitro fertilization, William May says:

> *"In Vitro fertilization makes it possible for Human life to be conceived outside the body of the (genetic) mother, but it is still a form of generating human life that is genetic, ie, possible only by uniting a male gametic cell, the sperm, with a female gametic cell, the ovum."*[26]

Here William May explains that there is no physical contact of the couple or persons in sexual act. The process is carried out in the laboratory where human life is manufactured in a Petri dish. The gametic materials required are just obtained from the individuals while the scientists carry out the process.

IVF serves many purposes as the proponents explained. It may be used to overcome female infertility in the woman due to faults in fallopian tube, which makes in vivo fertilization difficult. It could also be a relief for males who have defects in

---

[25] Assisted Reproductive technologies, in vitro fertilization in http://novlm.hoo.net...len.wikipedia.org/wiki/in vitro_ fertilization.
[26] William May, Catholic Bioethics and the Gift of Human Life "p.77

sperm quality, and in such cases, intacytoplasmic sperm injection (ICSI)[27] may be used where a sperm cell is injected directly into the egg cell. Here also the sperm have difficulty penetrating the egg and, and in these cases the partner's or donor's sperm may be used. This presents us with the idea of "homologous and "heterologous" in vitro fertilization.

1.1.1 HOMOLOGOUS IN VITRO FERTILIZATION

Here the gamatic materials used for the IVF are donated by the couples themselves. The husband's sperm is collected while the wife's egg is retrieved for the laboratory fertilization. Thomas O'Donnell explains that:

> *"Homologous IVF and ET is brought about outside the bodies of the couple through actions of third parties whose competence and technical activity determine the success of the procedure"*[28].

Here the husband's sperm is collected (through masturbation or the use of perforated condom)[29] while the wife's egg is

---

[27] http://www.motherjones.com/news/feature/2006/07/souls_on_ice.html
[28] Thomas J. O'Donnell, S.J Medicine and Christian Morality, P.507

retrieved but conception is performed in the laboratory." Conception "in vitro" is the result of the technical action which presides over fertilization.[30]" In homologous in vitro fertilization, the generation of the human person lacks proper perfection, that is, it is not through conjugal act, in which couples cooperate with God for the life giving to a new person.[31]

1.1.2 HETEROLOGOUS IN VITRO FERTILIZATION

This is different from homologous in vitro fertilization in the sense that the gametic materials are not exclusively obtained from both parents. There is a donor outside the couple. The commonly donated material is the sperm. Here the Ova donation may be from the wife while sperm is donated by another man to fertilize the egg. The husband's sperm may be used while egg is donated by another woman other than the wife. After fertilizing the egg, the resultant embryo is planted in the womb of the woman (not the one who donated the egg). The woman who carries this embryo is called, 'surrogate' mother"[32]. The embryo is transferred using a laparascope into the fallopian tube.[33]

---

[29] William May, op cit, P.75
[30] Thomas J.O'Donnell, S.J, p cit, P.507
[31] Pope John Paul II, Apostolic exhortation "familiaris consortio", no 14
[32] William May, p.78.
[33] Ibid

1.2 ARTIFICIAL INSEMINATION

This can be described as

> *"A process by which the male spermatozoa and the female ovum are brought together for achieving conception apart from and wholly distinct from an act of marital intercourse, whether using the spermatozoa of a third party (Artificial Insemination by Donor (AID) or the collected spermatozoa of the proper spouse (Artificial Insemination by Husband (AIH).*[34]

What Thomas O'Donnell emphasizes here is that this act differs from the normal process of conception where the husband has sexual intercourse with the wife otherwise, artificial insemination is defined as:

> "The process by which sperm is placed into *the reproductive tract of a female*

---
[34] Thomas J. O'Donnell, SJ, p.260

> *for the purpose of impregnating the female by using means other than sexual intercourse."*[35]

Here also, sperm is collected from the husband or donor (and has to be specifically freshly ejaculated sperm) after which it is frozen and thawed (in case sperm is from sperm bank) and injected into the female fallopian tube with intention of fertilizing the egg. The sperm could be collected through masturbation electric stimulator or use of perforated condom. Fertilization here is inside the woman after the sperm is injected through means other than coitus. Artificial insemination is used primarily to treat infertility using sperm injection. The sperm is placed in the cervix or after "washing", into the female's uterus by artificial means.[36]

This procedure requires that both the ovulation and the menstrual cycles of the woman be closely observed. When an ovum is released, the provided sperm is inserted into the Woman's vagina or uterus. In the case of vaginal artificial insemination, semen is usually placed in the vagina using a needleless syringe.[37] A longer tube, known as 'tom cat' may be

---

[35] http://www.webmd.com/infertility.and-reproduction/guide/semenanalysis
[36] Ibid
[37] Ibid

attached to the end of this syringe so that sperm is deposited deeper inside the vagina[17]. Also a specially designed cervical cap, a conception device may be used. When the procedure is successful, pregnancy results. Pregnancy from artificial insemination is the same as pregnancy carried by a woman impregnated through sexual intercourse. The difference is the procedure. Artificial insemination may be homologous or heterologous.

2.2.1 HOMOLOGOUS ARTIFICIAL INSEMINATION

This is a case where the sperm is from the husband. Here the husband's sperm is collected through masturbation or electric stimulation. It can also be collected through sexual intercourse where the husband wears a perforated condom. This is used if the man has no genetic disorder. If the husband is not able to ejaculate within the vagina, the sperm could be obtained through surgical removal of sperm from the epidermis, where sperm is stored.[38] The sperm may be washed in a laboratory "to remove antibodies and prostaglandins and to capacitate the sperm for fertilizing the ovum."[39] The husband's sperm could also be preserved or stored using the

---
[38] William May, Op Cit, p. 75
[39] Ibid, p. 76

cryopreservation of sperm in case of the husband's death so that the widow could have a baby afterward.

2.2.2 HETEROLOGOUS ARTIFICIAL INSEMINATION

Unlike the homologous, heterogonous Artificial Insemination is the case where the sperm is provided by a donor. The sperm is not from the husband but may be acquired from a sperm bank or another man.

Here the child to be born will have no genetic similarity with the husband which makes their child genetically incompatible. Lesbians and surrogate mothers use this procedure to acquire pregnancies. The danger here is the proficiency of the sperm and whether the sperm is disease free. This kind of practice puts a question not only on human procreation but also the marital union of the couple who supposed to be one body and one flesh. Masturbation puts a question on the essence of the conjugal act.

2.3 SEX SELECTION

This is a new technology which allows a couple to choose the sex of their child before pregnancy. This is otherwise known as 'gender selection.

> *"Every egg contains one X chromosome while sperm contains either an X or a Y chromosome. When an X-bearing sperm fertilizes an egg, a girl is conceived, and when a Y-bearing sperm fertilizes an egg, a boy is conceived."*[40]

This means that couples now have the chance of deciding whether to have a baby boy or baby girl. This is a technique that favours a male child against a female or vice versa.

There have been some reasons why this technique is used. In the first place there is the factor of genetic disease which the couples are afraid of transmitting to their offspring. Genetic disorders associated with male children, such as henophilia and muscular dystrophy[41] are avoided by selecting a girl. There is the issue of family balancing. Here couples who often have a particular gender of children are faced with choice

---
[40] www.charedjourney.com/ivf/sex-selection
[41] Ibid

of having children of the opposite sex. The third factor is that of having another child of the same gender.

Three major techniques for sex selection include: Gradient method, flow cytometry and pre-implantation genetic diagnosis (PGO), in Gradient method, sperm from the father is placed in a rapidly spinning machine called a centrifuge. While spinning, the machine helps to separate sperm with Y-chromosomes from those with X-chromosomes, which are heavier due to more genetic material. The desired chromosomes are selected from the sperm and used in IVF in order produce the desired sex.

Flow cytometry uses fluorescent dye to highlight sperm that carry X chromosomes. This fluorescent dye adheres to genetic material within the sperm. Because X-bearing sperm contain more genetic materials, these sperm pick up more dye than the Y-bearing sperm. Another machine which is laser is used to separate the two types of sperm. The sperm with the appropriate chromosomes are then used in IVI or IVF. This technique has high success rate.

One of the most successful technique or method used in sex selection is the pre-implantation Genetic Diagnosis (PGD).

Here embryos are created in a laboratory and allowed to divide. After three days, one cell from each dividing embryo is removed and analyzed of DNA and genetic material. After determining the sex of the embryos only those embryos of the desired sex are implanted into the mother's uterus through IVF.[42]

2.4 GAMETE INTRAFALLOPIAN TUBE TRANSFER

This is a process where eggs are removed from a woman's ovaries, and placed in one of the fallopian tube along with the sperm. This is different from the IVF because fertilization is in the body instead of Petri dishes. In gamete intrafallopian transfer, it takes an average of four to six weeks to complete a process. The woman, who must take a fertility drug to stimulate egg production in the ovaries, is monitored by the doctor. The growth of the ovarian follicles is closely observed.[43] When they mature the woman is injected with Human chrolonic gonadotropin.[44]

After, the eggs are harvested approximately 36 hours later, mixed with the man's sperm, and placed back into the woman's fallopian tubes using a laparoscope.

---

[42] Dr. Richard Paulson (2007) Assisted Reproductive technology. http//www.videojug.com/interview/assisted-reproductive-technology
[43] Ibid
[44] Ibid

Gamete intrafallopian tube transfer is used in cases of fertility problem which relates to sperm dysfunction, and where the couple has idiopathic infertility.[45] The cause of this problem is unknown.

All these processes of artificial reproduction are geared toward manufacturing human life using laboratory equipments. Here human biotechnologists – now become life manufacturers instead of custodians. The question is how normal the children born through these processes will be both physically, mentally, socially etc.

---

[45] Ibid

While that amendment failed, human cloning continues to advance and the breakthrough in this unethical and morally questionable science is around the corner.

Mike Pence

CHAPTER THREE

3.0 CLONING OR AGAMETIC REPRODUCTION

3.1 OVERVIEW OF CLONING

The term, "clone" is derived from the Greek word, "Klon" which has to do with asexual reproduction, or as commonly known, vegetative reproduction. Cloning, etymologically denotes producing a group of identical entities or organisms; or in recent times, a clone is seen as organism that is a genetic copy of an existing organism.

The American Encyclopedia describes clone as a protocopy of an organism with the same genetic information. Thus it defines cloning as "the production of a group of genetically identical cells or organisms, all descended from a single individual."[46] This implies that the members of a clone have exactly the same characteristics and identity unless there is an environmental or mutational cause where development

---

[46] Academic American Encyclopedia, Grolier incorporated, Danbury, Connecticut, Vol. 5, p. 64

variability has to occur. Furthermore, "cloning is the process of creating an exact copy of a single gene, cell or organism."[47]

Cloning can also take the form of natural development of embryo in the case of twin babies. Here there is no use of other body cells other than gametes (ones with the nucleus, which contains the genetic material, that is the gametic cells from the organism being cloned).[48] It takes a natural process here involving the egg and sperm.

The Academic American Encyclopedia also observes that in species whose reproduction is strictly asexual, each population consists of one or more clones, depending on the number of ancestral individuals. "Such species include all bacteria and blue-green algae, mostly protozoan and other algae, some yeast, and even some higher plants and animals, such as dandelions and flatworms.[49] This ascertains cell divisions in the reproductive systems of some plants and animals which are asexually cloned. It is unlike sexual reproduction where clones are created when a fertilized egg splits to produce identical (monozygous) twins with identical DNA.[50]

---

[47] Microsoft Encarta, 2009, from 1993-2008 Microsoft Corporation, p. 1
[48] National Catholic Reporter on Cloning, Oct 22, 1999
[49] Academic American Encyclopedia, Op Cit, p. 64
[50] A Press release, by US General Assembly adopting United Nations Declaration on Human

Cloning occurs naturally and can occur in organisms that reproduce asexually as well as those that reproduce sexually. While that of sexual reproduction is a relatively rare event, many species produce their descendants asexually without the combining of the male and female genetic materials that occur in sexual reproduction. Such offsprings are clones of the parents.[51] These are obtained through "molecular" or "cellular" cloning.

Molecular cloning determines the identify of each cell or molecules to others. This has to do with the DNA (deoxyribonucleic acid). The molecular basis of genes, is a fairly routine occurrence for the molecular biologists.

> "DNA fragments containing genes are copies and amplified in a host cells usually is bacterium. There are many scientific experiments that rely on the availability of large quantities of DNA molecules. Molecular cloning is the mainstay of recombinant DNA

Cloning,
   8th March, 2005
[51] Ibid

> technology. It has led to the production of many important medicines, such as insulin to treat diabetes and tissue plasminogen activator (TPA) to dissolve clots after a heart attack".[52]

This recombinant DNA formation is also known as gene splicing. It is a procedure whereby segments of genetic material from one organism are transferred to another. The basis of this technique lies in the use of special Enzymes that split DNA strands where certain sequences of nucleotides occur in.[53] There are series of donor DNA fragments from other organisms. These DNA fragments are, in some places, combined with virus or with plasmids that is, small rings of self-replicating DNA found within cells.

> "The virus or plasmid vectors carry the donor DNA fragments into cells. The combined vector and donors DNA fragments constitute the recombinant DNA molecule. Once inside a cell, referred to as a host, this molecule is

---

[52] Appleyard, Bryan, "Brave New World: Studying Human in the Genetic Future." Viking Press, New York, 1998
[53] Academic American Encyclopedia, Vol. 9, p. 84

replicated along with the host's DNA each time the host divides. These divisions produce a clone of identical cells, each having a copy of the recombinant DNA molecule and the potential to translate the donor DNA fragment into the protein it encodes."[54]

This shows molecular cloning as natural cloning. It equally shows that the process of reproduction in asexual reproductions take a natural process. It is not manipulated by any artificial scientific interruption.

But in cellular cloning, cells are grown in a culture in the laboratory to make copies of them and these are derived from the soma, the body. The genetic make up of the resulting cloned cells, called, "cell line" are identical to the original.[55]

This procedure, like molecular cloning, is also very useful in the testing and sometimes the production of medicines, such as insulin and EPA. Thus, molecular and cellular cloning of this

---

[54] Ibid
[55] Annas, George J. "Regulatory Models for Human Embryo Cloning: The Free Market, Professional
  Guideline, and Government Restrictions" Kenedy Institute of Ethics Journal 4, 1994, p. 235-249

sort do not involve germ cells (egg and sperm); therefore, they are not capable of producing a baby.

On the other hand, natural cloning occurs in the development of identical twins. Thus, identical twins are example of how clones don't have to start in a test tube in the laboratory. Naturally, twins occur just after normal fertilization of an egg cell by a sperm cell in a woman's womb. Though, in rare cases, the resulting fertilized egg (zygote) tries to divide into a two-celled embryo, the two cells separate. Here each cell continues dividing on its own, ultimately developing into a separate individual within the womb. Since the cells come from the same zygote, the resulting individuals are generally identical.[56]

> "Artificial embryo twining uses the same approach, but it occurs in a Petri dish instead of in the mother's body. This is accomplished by manually separating a very early embryo into individual cells, and then allowing each cell to divided and develop on its own. The resulting embryo are placed into a

---

[56] Article on Human Cloning, http//www.clhr-chawks.org/.../natural-cloning

> *surrogate mother, where they are carried to term and delivered...they are genetically identical".*[57]

We will treat artificial cloning in human in details, but here we focus on the natural process of developing twin babies.

> *"Identical twins are in effect clones. They both have the same genetic material. But they do not have the same experiences, nor do they have the same souls. They are different people...so a clone would be an identical twin, not an exact copy of the person which includes experiences and memories. The clone becomes a new person."*[58]

This whole thing about cloning of twins is an assertion of the fact that naturally organisms clone themselves but the question is why do scientists insist in artificial cloning of animals, and human beings in particular?

---

[57] Ibid
[58] Investigative reports on identical twins>http://www.boards.actv.com/.../231657.

3.2   ANIMAL CLONING

Animal cloning has been the subject of scientific experiments for years, but garnered little attention until the birth of the first cloned mammal in 1997, a sheep named, "Dolly".[59] But since the production of Dolly, several scientists have cloned other animals, including cows and mice. The recent success in cloning animals has sparked fierce debates among scientists, politicians and the general public about the use and morality of cloning plants, animals and possibly humans.

But in the first place, we ask what is animal cloning and what are the techniques used to clone animals?

> *"Animal cloning is the process by which an entire organism is reproduced from a single cell taken from the parent organism and in a genetically identical manner. This means the cloud animal is an exact duplicate in every way of its parent; it ahs the same exact DNA."*[60]

---

[59] <http://www.science.howstuffworks.com/genetic-s...
[60] Stephen Ferry <www.buzzle.com/.../animal-cloning

With the development of bio-technology, animals are now being cloned artificially. This took a very long time and many attempts were made which had proved abortive. The first fairly successful result in cloning animals artificially was the cloning of a tadpole from a frog's embryonic cells.[61] This was done through the technique called, "nuclear transfer". The tadpole so created did not survive to grow into natural frogs, yet is was a very successful breakthrough.

The first successful mammal cloned was Dolly the sheep, who not only lived but went on to reproduce herself and naturally. The creation of Dolly was by Ian Wilmut and his team at the Roslin Institute in Edinburgh, Scotland, in 1997.[62] Since Dolly, "scientists have been successful in producing a variety of other animals like: rats, cats, horses, bullocks, pigs, deer, etc."[63] The question now is how did scientists finally succeed in artificially cloning these animals?

In the cloning of animals, three categorical methods are involved, namely:

1. "Blastomere separation,
2. Blastocyst division (twinning), and
3. Nuclear transfer."[64]

---

[61] Ibid
[62] Ibid
[63] Ibid
[64] "Firm says It created Embryo Out of Human, Cow cells" By Rick Weiss

"Blastomere" separation involves the splitting of the embryo immediately after fertilization. Each of the separated cell is called a "Blastomere" and is able to produce a new individual organism.[65] These "Blastomere" are totipotent which shows they have the potentials of producing an entirely new organism. Also this allows animal embryos to be split into several cells to produce multiple organisms of identical genetic make-ups.[66] Scientists use this in livestock breeding.

On the other hand, blastocyst division or twining involves a process whereby an embryo that has already been formed sexually is split into two identical halves. These two parts is now transferred to the uterus.[67]

The last of the categories is "Nuclear Transfer". Here the DNA nucleus extracted from an embryonic cell is implanted into an unfertilized egg, from which the existing nucleus had already been removed. The membranes of the "blastomere" and the enucleated eggs are fussed together artificially. Thus, the

---

<http://www.washintonpost-com/wp-s...tional/scrence/cloning/cloning.htm>Nov.13, 18\998
[65] Ibid
[66] Ibid
[67] Ibid

nucleus from the "blastomere" enters the egg cytoplasm and directs development of the embryo.[68]

But in the production of Dolly the sheep, the Roslin institute used a variant of nuclear transplantation. In this process,

> *"The nucleus that programmed the creation of Dolly was transferred from the adult sheep mammary cell, not from an embryo...unfertilized egg being removed and replaced with a nucleus obtained from a specialized cell of an adult (or fetal) organism. Since almost all the hereditary material of a cell is contained in the nucleus, the enucleated egg and the individual into which that egg develops are genetically identical to the organism that was the source of the transferred nucleus."*[69]

---

[68] Stephen Ferry <http://www.buzzle.com/.../animal-cloning
[69] "Hello, dolly, Dolly, Dolly..." >http://www.nytimes.com/books/98109/06/reviews/980906.06 papinet.html

The cloned animal that resulted from this process had a genetic make-up exactly identical to the genetic make-up of the original cell. Animal cloning today can be done both for reproductive and non-reproductive purposes. Cloning here is done to produce stem cells or other such cell that can be used for therapeutic purposes. This will be discussed in details. But while we consider the quick progress in animal cloning, we have to weigh the benefits of this activity and see how comfortable the idea of animal cloning is.

3.2.1 THE BENEFITS OF ANIMAL CLONING

It is clear that there must be those who embrace the pros of animal cloning more than the cons. Those who support animal cloning do so on the basis of possible benefits while those who condemn do so on the basis of ethics. Some people argue that it is against nature and ethically wrong, mostly because the scientists use the idea to dive into human cloning.

It could be said that in a large percentage of cases, the process of cloning animals fails in the course of pregnancy or some sort of defects occurs, like in a case of "a calf born with two faces. Sometimes the defects manifest themselves later and kill the clone".[70] This is to say that there are defects in most cases which defeats the idea of cloning the animal. But this may not be enough reason to reject cloning of animals. There are many benefits which animal cloning hold that prompt much support.

> "On the favourable side with successful animal cloning – particularly cloning from adult animal – you know exactly how your clone is going to turn out. This becomes especially useful when the whole intention behind cloning is to save a certain endangered species from becoming totally extinct."[71]

This is to say that one of the benefits of cloning animals is to save a possible extinct of certain species of animals. Animal cloning equally helps in boosting and improving livestock farming. It is on record that livestock cloning is the most recent evolution of selective assisted breeding in animal husbandry given the fact that formally, artificial insemination, in vitro fertilization, embryo transfer were used by farmers, ranchers and pet enthusiasts with powerful tools for breeding the best animals.[72]

---

[70] Animal Cloning>http://www.buzzle.com/.../animal-cloning
[71] Ibid

"Cloning animals is a reliable way of maintaining high quality and healthy livestock to supply our nutritional needs and consumer demand. Identifying and reproducing superior livestock ensures herds are maintained at the highest quality possible. Animal clones will primarily be used as breeding stock to improve the health and quality of animal used for food production".[73]

It is good that animal cloning helps improvement of livestock but will most consumers like to eat cloned animals? Also Ethically and morally, animal cloning may not pose any threat but another question is why scientists are too keen in cloning human beings as well? But these question not withstanding animal cloning is a welcome activity since it can improve human life, tempering on human dignity.

3.3 HUMAN CLONING

---

[72] Cloning Fact Sheet>http://www.cra:gov/hymis/elsi/cloning.shtml
[73] Ibid

Human cloning is the application of the cloning process to the reproduction of human beings. In human cloning, human becomes both the specimen and the targeted result. This is in the sense that it is the human body materials that are being used for the process and the result is also to get a specific human being or human materials. Thus human cloning targets the human being without much question on his intrinsic nature but focuses more on the extrinsic gain.

Also on the nature of human cloning, Peter J. Russel, says that "Human cloning is the creation of a genetically identical copy of an existing or previously existing human being."[74] It is the creation of a genetically identical copy of a human cell or human tissue.

> "The term is generally used to refer to artificial human cloning; human clones in the form of identical twins are commonplace, with their cloning occurring during the natural process of reproduction."[75]

---

[74] Peter J. Russel, "Genetics: A Molecular Approach", San Francisco, California, USA, Peatson
    Education, 2005
[75] Article on Human Cloning >en.wikipedia.org/wiki/Human-cloning.

This shows that in artificial human cloning, a human individual who wants to have a replica of himself or herself, uses artificial means to "photocopy" self without undergoing the natural process of childbirth. Here a woman may not need a man's seed in order to reproduce since she can make use of her adult cell to fertilize her enucleated egg, like-wise a man using an egg whose nucleus is removed to produce a child without undergoing the stress of sexual intercourse. The whole process lies on the availability of adequate technology which can give rise to the intended result.

Also human cloning could refer to the creation of embryos through somatic cell nuclear transfer (SCNT). Here the intention is not to produce offspring but for use as a scientific tool or materials for medical ends. Such process or form of cloning gives rise to what could be referred to as "research cloning" or "therapeutic cloning." These will be discussed in details. Basically, human cloning will also take the process which saw the successful cloning of Dolly the sheep. Thus, if the scientist will ever clone human beings it will be through "reproductive cloning."

Human cloning involves three major types: Therapeutic cloning, reproductive cloning and replacement cloning.

3.3.1 HUMAN THERAPEUTIC CLONING

Therapeutic cloning involves cloning for the sake of medicine. It is cloning human embryos in order to harvest the stem cells for medical use or for research purposes. It is also called "embryo cloning", that is, the production of human embryos for use in research.

> *"The goal of this process is not to create cloned human beings, but rather to harvest stem cells that can be used to study human development and to treat disease. Stem cells are important to biomedical researchers because they can be used to generate virtually any type of specialized cell in the human body."*[76]

What this claim shows is that cloning human organs help to reduce disease by planting a new organ from the clone of the same individual, which makes it possible for the body to accept

---

[76] Watson, James, "Moving Toward a Clonal Man: Is This What we want?" The Atlantic Monthly 1971

the treatment. Though there are complications here since it involves other complexities, including the claim by being some scientists that immune rejection still exists in therapeutic cloning. Other complications of therapeutic cloning involves the method and process of this cloning.

"In therapeutic cloning, stem cells are created from a donor for the main purpose of providing tissues (such as for organ repair), in the event that the donor might need such treatment at a future date."[77] The tissues and organs are cloned and preserved waiting for a future disease which may call for organ transplantation of the donor.

Therapeutic cloning also involves the transfer of a somatic (adult) cell from the donor into an enucleated egg, thereby producing a single celled cloned embryo.[78] The egg, when developed into a blastocyst, the inner cell mass is then removed and cultured into embryonic stem cells, which are grown to produce the desired healthy "therapeutic cells."[79] These cells include: nerve cells, muscle cells, organ tissues etc. The new cells are transplanted to the patient, who have donated the original somatic cell.[80]

---

[77] Therapeutic Cloning of Stem Cells <www.cellmedicine.com/cloning.asp
[78] Ibid
[79] Ibid
[80] Ibid

While adult stem cell cloning involve cloning cells from tissues obtained from different parts of the human body, like bone marrow, blood vessel, skeleton muscles, umbilical chord, brain, amniotic fluid during pregnancy, placenta after pregnancy, there is one form which deals with cells from embryos of unborn babies. Though in therapeutic cloning, the goal is producing stem cells for cure.

Stem cells are extracted from the egg after it has divided for five days. This process destroys the embryo, and this has raised series of ethical concern. The proponents of therapeutic cloning hopes that one day stem cells can be used to serve as replacement cells to treat heart disease, Alzheimer's cancer, and many other diseases.[81]

Originally it looked more promising using embryonic stem cells obtained through IVF because of its newness and lively nature. They are innocent and unadulterated, and undifferentiated. Until about 14th day of existence, after fertilization, the embryonic cells organize themselves to become specific tissues and organ in themselves. This is referred to as a process of "differentiation".[82] They seem to be

---

[81] "Campaigners Win Cloning Challenge" BBC News 15 Nov. 2001
http://news.bbc.co.uk/1/hi/sci/tech/1657707.stm

capable for genetic manipulations. Embryos created through IVF could be implanted before 14 days of existence so that the parts will not start to be shooting out already.[83] This means that in therapeutic cloning, the embryos from IVF could be used culturing it early to, form human organs. They are more promising than embryos cloned through agametic processes. In IVF, fertilization takes natural process involving the gametes, and such embryo will be more productive in experiment.

3.3.2 HUMAN REPRODUCTIVE CLONING

In reproductive cloning, the intention is to reproduce an exact person who has the same genetic information with the individual cloning himself/herself. A cloned human being here will be a prototype or a photocopy of the individual who is cloning himself/herself.

"Reproductive cloning is a technology used to generate an animal that has the same nuclear DNA as another currently or previously existing animal."[84] This was the method that

---

[82] Codification Division, Office of Legal Affairs, United Nations (18 May, 2005). "Ad Hoc Committee on an international convention against the Reproductive Cloning of Human Beings."
United Nations. http:www.un.org/law/cloning/.
[83] Ibid
[84] Ibid

generated Dolly the sheep. It takes the process of somatic cell nuclear transfer.

> *"In a process called "somatic cell nuclear transfer" (SCNT), scientists transfer genetic material from the nucleus of a donor adult cell to an egg whose nucleus, and thus genetic material, has been removed. The reconstructed egg containing the DNA from a donor cell must be treated with chemicals or electric current in order to stimulate cell division. Once the cloned embryo reaches a suitable stage, it is transferred to the uterus of a female host where it continues to develop until birth."[85]*

Animals or possibly human being could be created using nuclear transfer technology. But these clones may not truly be identical to the original animal or human being as the case may be. "Only the clone's chromosomal or nuclear DNA is the same as the donor. Some of the clone's genetic materials

---

[85] Ibid

come from the mitochondria in cytoplasm of the enucleated egg."[86]

In somatic cell nuclear transfer, human begins could be produced who have the genetic material of only one parent. What scientists do here is to transfer the genetic material from a donor's somatic cell to an enucleated egg cell. The enucleated egg posses no genetic material because it has already been removed. The human being who will result here will have the genetic information of the person with whose somatic cell he or she is cloned.

This procedure is carried out by merging the somatic cell and enucleated egg cell using fusion or injection.[87] The fusion material makes the patient's immune system, in therapeutic cloning, not to reject the stem cell as foreign material. In reproductive cloning, nuclear replacement could also be used. Nuclear replacement involves the introduction of genetic material into the cytoplasm of an unfertilized egg or embryo, whose genetic material or nucleus has been removed. This genetic material is in form of an individual cell.[88] In this technique, the nuclear genes of clones that are produced would

---

[86] Ibid
[87] Microsoft Encarta, 2009 called from 1993-2008 Microsoft Corporation
[88] FAQ>www.wikipedia.org/wiki/Human-cloning

be identical, although the mitochondria DNA of those clones would be different.[89] Unlike the embryo splitting technique nuclear replacement has the ability to create a clone of an adult organism as well as the potential to produce many more clones.

Nuclear replacement is a relatively new technique. It was first used in 1952 in creating frogs.[90] But the more recent developments of nuclear replacement technology have brought with them the potential to contribute towards genetic improvement of livestock. Since the sheep Dolly was produced in 1997, this technique has reportedly been gaining more ground towards cloning human beings.

3.3.3 EMBRYONIC STEM CELL HARVESTING

This has been the most difficult and controversial aspect of cloning in the sense that it has given rise to more religious and ethical questions. The researchers clone human embryos in order to harvest the stem cells thereby destroying the embryos. This has ethical significance because the new human life formed in the embryo is destroyed. Religious people do not accept this since it not only dehumanizes human beings but

---

[89] Ibid
[90] Ibid

also violates the sacredness of the human life. But before evaluating this technology, let us briefly discuss the nature of stem cells.

Stem cells are those "cells that have the ability to divide indefinitely and to give rise to specialized cells as well as to new stem cells with identical potential."[91] These cells have the capacity to renew themselves both in the "undifferentiated states as well as differentiate into descendent cells that have a specific function."[92] The stem cells include cells from the skin, blood, uterus and cells that line the gastrointestinal.[93] The body organs renew themselves periodically because of the presence of the stem cells. Those parts of the body that lack the stem cells are not capable of renewal. For instance, the heart do not have stem cell, and once dead, cannot be renewed. But the embryonic stem cells are different from adult stem cells. The embryonic stem cells are capable of generating, virtually, all types of cells in the *foetus* and even in the mature body. Thus, the embryonic stem cells are *pluri-potent*.[94]

> *"Pluripotent cells, present in the early stages of embryo development, that*

---

[91] Justine N. Ekennia, Bio-Medical Ethics, Op cit, p. 279
[92] Ibid, p. 123
[93] Ibid
[94] Ibid, p. 124

can generate all cell types in a fetus and in the adult and that are capable of self-renewal. Pluripotent cells are not capable of developing into an entire organism."[95]

This is the reason why scientists prefer the embryonic stem cells to the adult/somatic cells. They say that the embryonic stem cell is more adventorious as it offers the possibility of generating all parts of the human organ. Stem cell harvesting is imployed in therapeutic cloning. They cannot develop into human beings or animals since after the extraction, the embryo is destroyed and dies off. The stem cells are not zygotes that develop into babies but could be cultured to develop various organs of the human body.

The proponents of therapeutic cloning have argued that stem cell harvesting is more beneficial to man. They claim to produce human parts through stem cell harvesting in order to cure patients who are in pain and in need of healing and cure.

"The first possible benefit of therapeutic cloning would be to

---
[95] Ibid, p. 278

produce human spare parts for transplants when needed. We are aware of the fact that human organs for transplantation are scarce resources...Hence the desire to clone an organ to replace a defective organ is the argument of pro therapeutic cloning scientists."[96]

Thus the scientists claim that by injecting cloned healthy cell into damaged heart tissues, disease like heart attack are likely to be reversed. "Some echo that with cloning, especially stem cell harvesting, Alzheimes, Parkinson degenerative joint diseases may be curable".[97] Also the scientists claim that stem cell research may also hold the key to slowing down the aging process.[98]

3.3.4 CLONING AND SURROGATE MOTHERHOOD

Surrogate mother-hood has to do with infertile women or wives who are willing to receive embryo from spermatozoa either from their husband or another man to be pregnant. The sperm is mainly obtained through masturbation and

---

[96] Ibid, p. 126-127
[97] Ibid, p, 127
[98] "Stem cells hold promise of cure" The National Catholic Reporter, October 22, 1999

subsequent adulterous artificial insemination.[99] The surrogate mother is willing to bring on the child and may be paid for the service (if not the wife of the husband).

In the case of cloned embryo, the surrogate mother carries the embryo for the husband whose genetic material has been cloned. The child to be born is clone of the husband with whose sperm the embryo was produced. In the case of a hired surrogate mother, she becomes richer with the money paid for the service while the infertile wife of the man has the baby.

Proponents of reproductive cloning aim at producing a complete and super human being who will have high intelligence, immunity and must efficient power. The embryo cloned in the laboratory is thus implanted into a surrogate mother who willingly accepts the service of carrying this baby to maturity and give birth to it at the duration of the pregnancy. The husband of an infertile wife paying off the surrogate mother receives the baby who is his other self, and his infertile wife take the responsibility of bring up her husband's baby in the pretence that she is the mother of the baby.

3.4 SCIENTIFIC AND MEDICAL ASPECTS OF HUMAN CLONING

---

[99] Thomas J. O'Donnell, SJ. Op Cit. p, 264

The tremendous progress made by science in the area of improving human health is not only undisputable but also commendable. It is clear that medical research is "patient centred". This is why the biotechnologists embark on therapeutic cloning and embryo research to solve the various problems posed by nagging disease. But it is important the personhood of man is taken into consideration. Though the patient has the right to care and medicine, he/she has the foremost right to life. In view of researching for a medical cure for patients with certain diseases, the question is whether the bio-scientists aim at proper means for the cure? The patient is a person who need care and cure, but the question is whether these cure and care should be at the detriment of another unique human being who also has the right to life.

It is obvious that science which approves of cloning for the sake of curing diseases are separating the ethical values from medical values and religious values. Though bio-technological sciences have medical benefits for the human person and his health, the question of values should not be compromised;

> *"Ethics in health care resides as we have seen, in human values, morals, cultures, intense personal beliefs and faith that have been individually shaped by events of our personal experiences in life that are embedded in the individual traditions of our communities and people. No one would be in doubt as to how ethics can enter into biomedical sciences. But issues of values are the most problematic in the history of ethics."[100]*

This means that the scientist who aims at cloning human embryo for therapy has to weigh the ethical values of his researches. There must be an underlined respect on the status of the human embryo and the human zygote. There must be an answer to when the human life actually begins. But to take care of these problems, the real meaning of human life must be put clear.

According to Justin N. Ekennia, "Biologically, a human being is the life that emanates from human conception; that is,

---

[100] Justin N. Ekennia, Op cit, p. 11

the fusion of a male sperm or adult DNA with a female egg (ovum).[101] The fact is that human life begins with the amalgamation of the sperm and the egg. But life can also emerge without the sperm. This means that the somatic or adult cell could be used, as in agametic reproduction, to fertilize an egg.

> *"It must be added that conception can also take place without the male sperm coming in contact with the female egg. The recent technology of human cloning with Adult DNA fertilizing a female egg, has shown that human beings can result out of cloning. Or even a woman's egg can be acted upon through chemical processes to fertilize itself, a process called parthenogenesis."*[102]

The main issue here is the beginning of life. Whether it is form from the egg and sperm, adult cell and egg or itself with chemical process as in parthenogenesis, the issue is that life is formed immediately there is fertilization. Thus, the question of

---
[101] Ibid, p. 13
[102] Ibid, p. 40

harvesting stem cells from the embryo for therapy remains quite controversial because the scientist neglects the new life which is formed in order to rescue a sick patient. Since the new formed life in the embryo is a unique life itself, the question then is why the new life should be destroyed in order to sustain another life.

Human cloning thus poses a very vital question in the area of medicine. The health of a human person who needs care must be next to the life of an embryo who is a unique but dependable human being since it is still fragile, vulnerable and cannot be on its own. This means that the human being who is still in form of embryo is still fragile and vulnerable and must be protected as a patient who is dependable and needs care and cure. The life of the embryo must not be sacrificed for the life of the sick individual. Both lives are important but more emphasis should be placed on the life of the embryo.

Suppose That Every Prospective Parent In The World Stopped Having Children Naturally, And Instead Produced Clones Of Themselves. What Would The World Be Like In Another 20 Or 30 Years? The Answer Is: Much Like Today. Cloning Would Only Copy The Genetic Aspects Of People Who Are Already Here.

Nathan Myhrvold

CHAPTER FOUR

4.0 ETHICAL EVALUATION OF HUMAN CLONING

4.1 THE CATHOLIC CHURCH'S PERSPECTIVE

In the first place, there have been some established facts in science about the reproduction of man from sexual union to fertilization and growth and development of the embryo. Though there are some who, out of selfish interests have put up some scientific statements that will suit their desires, there are many also who have laid down the true findings of their scientific researches without attaching subtle interests.

Concerning the development of the human embryo, the late Professor Jerome Le Jeune of Rene Descartes University of Paris, Faculty of Medicine was quoted as saying that:

"A new human life begins when a sperm fertilizes an egg forming a zygote. At the moment of conception, a unique genetic and personal constitution is spelled out for the specific human being created, whose personal constitution has not occurred before and will never occur again. The zygote is the most specialized cell under the sun in that no other cell will ever have the same instructions in the life of the individual being created."[103]

This implies that the human life which begins with fertilization has the constituents of a human being. The zygote has a soul since it is a unique human being and the life is also valued as a mature human life is valued. Whether the whole information of the human life in the embryo has been spelled out or not, "as soon as he has been conceived, a man is a man."[104] The Church is the protector of life. Human life is a precious gift of God that is all good. It is an act of God's beautiful and unfathomable love. The church teaches that life is

---

[103] Justin N. Ekennia, *Op Cit*, P. 265
[104] Ibid, P. 266

sacred and inviolable and must be respected, from birth to death.

The voice of the church concerning assisted reproductive technology in general and human cloning in particular could be heard through Papal encyclicals, exhortations, Papal conferences, address to the United Nations, various Catholic biomedical academies and some other avenues where the Church can air out her views on current issues in the society.

> *"The Church respects and supports scientific research when it has genuinely human orientation, avoiding any form of 'instrumentalizing' or destruction of the human being and keeping itself free from the slavery of political and economic interests".[105]*

What the Church is saying here is that science and its researches are good and acceptable if and only if they do not destroy life nor make life a tool or means to a certain end. The Church is saying that science should protect and promote life

---

[105] Proceedings of the Ninth General Assembly of the Pontifical Academy for Life: "The Ethics of
   Biomedical Research for a Christian Perspective" no. 2, Vatican, Feb 24-26, 2003

and not destroy it. If the research is for the benefit of humanity it means that such research cannot at the same time be a threat to life. The Holy See welcomes researches and investigations in the field of medicine and biology whose goal is curing of diseases and improving of life quality. The Holy See insists that such research must respect human dignity and must be consistent with this respect.[106]

According to Pope John Paul II,

> *"...the magisterium of the Church appeals to the universality and the dynamic and perfective character of the natural law when referring to the transmission of life, whether it be to maintain the fullness of the spousal union in the procreative acts and to preserve the openness to life in the conjugal act."*[107]

For the Church, the natural right which protects human dignity emanates from the very nature of man and ought not be the expression of individual interests of those who are able to participate in social life or those who hold the consensus of the

---

[106] Document of the Holy See on Human Cloning, P. 1
[107] John Paul II, Encyclical: "On Human Life (Humanae Vitae)", n. 10

majority.[108] The dignity of human life does not lie in the power of man but God. God gives man the dignity and integrity of life and no human being should dare take them away. It implies that human life must be esteemed at every stage beginning from conception till natural death. This is why Pope John Paul II rejects whatever opposes, violates, insults and is disgraceful to human life. For him, actions like murder, genocide, abortion, Euthanasia, willful self-destruction, mutilation, coercion, arbitrary imprisonment, slavery, prostitution are all infamies.[109]

Cardinal Cathal B. Daly says that the central citadel of Catholic morality is the doctrine of the natural law. And for him, it is of great importance today because it is "rational, personalistic morality" which stands in great contrast to the "inhuman and anti-personalistic morality of liberal agnosticism and scientific humanism."[110] This natural law is well defined in the Catholic Catechism. It states that everything is created for man to enable him realize his vocation, which means that man must not regard created goods as ends in themselves but must offer them back to serve God by using them to serve him.[111] This implies that scientific resources should not be ends to themselves but means to serve God.

---

[108] Justine N. Ekennia, Op Cit., P. 215
[109] Evangelum Vitae, no
[110] Cathal B. Daly, "Morals, Law, Life", Scepter Publisher, Chicago, 1966 P. 29-30
[111] CCC, no. 357

On the other hand, the Vatican II Fathers talk of "Divine Law". This law is the highest norm of life. It is eternal, objective and universal. Here God gives orders, directs and governs the entire universe, and all the ways of human community is planned in this wisdom and love. "Man was made by God to participate in this law, with the result, that, under the gentle disposition of Divine providence, he can come to perceive ever increasingly the unchanging truth."[112] This is the reason why man should recognize the voice of God which urges him towards goodness and to avoid what is evil. "Everyone is obliged to follow this law, which makes itself heard in the conscience and is fulfilled in the love of God and of neighbour."[113] Thus, whatever is holy and sacred should be valued as such. Also the Church asserts that God fashioned man is His own hand and impressed His form on him in such a way that even "what was visible might bear the divine form."[114] In this case man was made in the image and likeness of God, he bears the divine form. He is "the only creature on earth that God has willed for its own sake"[115] and he is called through knowledge and love to share in God's own life. "...he possesses the dignity

---

[112] Second Vatican Council: "Dogmatic Constitution on Dignity of the Human Person (Dignitatis Humanae)", Rome, no. 3
[113] Catechism of the Catholic Church (CCC), no. 1706
[114] CCC, no. 704
[115] CCC, no. 356

of a person who is not just 'something' but rather is 'someone'."[116] What is implied here is that man, from the stage of zygote is formed in the divine image and is someone not something. This makes assisted reproductive technologies, especially human cloning a moral evil.

For these reasons, the Roman Catholic Church opposes artificial reproductive technologies using human embryo. The Church opposes all kinds of in vitro fertilization because it separates the procreative purpose of the marital act from its unitive purpose. The reason is that the fundamental nature of martial act, while uniting husband and wife in the closest intimacy also renders them capable of generating new life. As a result, if each of these qualities (the unitive and procreative) is preserved, the marriage fully retains its sense of true mutual love which is the intention of God. As we are going to see later in this chapter, God has willed an inseparable connection in marriage, "which man on his own initiative may not break, between the unitive significance and the procreative significance which are both inherent to the marriage act."[117]

Generally speaking, in vitro fertilization departs from God's plan for human reproduction because it dispenses with

---

[116] Eamonn Keane, B. Comm. Dip. Ed Grad., "Human Life International, Australia Inc., P. 69
[117] Humanae Vitae, no. 12

the act of martial union. Cloning "represents a radical manipulation of the constitutional relationality and complementarity which are at the origin of human procreation,"[118] because it takes the additional step of dispensing with human genetic. Human cloning is immoral because it attacks the dignity of human procreation and because it is an "affront to the dignity of the individuals involved."[119] Human dignity is innate; bestowed upon us by God. It is not based on the ability to care for ourselves or competence to complete a task. Being dependent upon other does not cause us to lose our dignit."[120]

The Church condemns all activities using human embryo or human gametes (sperm and ova) with respect to artificial reproduction and therapeutic research.
"The Holy See opposed the cloning of human embryos for the purpose of destroying them in order to harvest their stem cell, because it is completely in compatible with the respect for the dignity of human beings."[121]

4.2 CLONING AND HUMAN SEXUALITY

Man as a social and sexual being is born with the desire for human relationships but the ability to satisfy this quest

---
[118] http://www.cuf.org/Human-cloning
[119] Ibid
[120] Arcanum No. 7
[121] Document of the Holy See on Human Cloning, P. 1

must be shaped and sculpted over a long process. Man naturally longs for intimacy (between the opposite sex) yet "to become intimate comes about only through a developmental process, a continuum of growth that begins before we are born and unfolds throughout our entire lives."[122] This is why those who did not go through the right process of sexual development have the tendency of neglecting the rightful attitude towards managing their sexuality.

According to Karl H. Peschke, "to be human is to be born of other men. There is a man and woman and a family behind every person."[123] This shows that to be a human being and live naturally as a human being, one must be first of all conceived and born as every other human being. The person must be born of a man and woman in a family and must be live through the natural process of human development. This is the act of being human, and it defines human sexuality.

"Sexuality is a fundamental component of personality, one of its modes of being, of manifestation of communicating with others, of feeling, of expressing and of living love."[124]

---

[122] Fran Ferder and John Heagle, "Your Sexual Self: Pathway To Authentic Intimacy, St. Paul's
Publications, Bandra Mumbai, 2001, P. 9
[123] Karl H. Peschke, Christian Ethics: Moral Theology in the Light of Vatican II Vol 2, Theological
Publications, Bangalore India, 2004, P. 418

Human sexuality is the integration of the whole being or beingness of man. It is living as God had created us. Human sexuality places much emphasis on the beginning of human life. This is because to be human is to be created and born through a natural human process of reproduction. The beginning of a new human life is very important because it is sacred. This is a time in which God's continuing creative power is especially evident. "A child is born in cooperation between parents and God."[125] This is a power naturally given to a man and a woman who have come together to live as husband and wife. They, through love and self-giving, give new life to a child through sexual intercourse. The process of living out this responsible cooperation, begins with conception and birth and is completed in the education of the child. God nourishes and guides the new human life through them.[126]

> "At the beginning of a new human life stands the affection and utter love of two people, father and mother. This love and its fruitfulness is based on the differences of sexes. The differences of sexes however are not only of

---

[124] Congregation for Catholic Education, "Educational Guidance in Human Love", Nov 1, 1983
[125] Karl H. Peschke, Op Cit., P. 419
[126] Ibid

> *significance in the creation of new life. People's sexual differentiation permeates their whole person always and forever; it stamps all their actions and is creative also under aspects other than procreation."*[127]

Children are the fruits of the relationship, love and intimacy which exist between the couple. But the couple only participate in creation as instruments. God is the creator. Therefore couples should not make themselves creators thereby playing God. They have no right also to resist reproduction during marital act. Human sexuality thus shows that the husband (male) with the wife (female) must live freely their human nature. The husband is attracted to the wife and vice versa. Each of them gives freely selfless love, intimacy and makes sacrifices which are necessary to their martial union. They should see sex as a means of integrating their love in marriage, aware that it can lead them to give life to a new human being. "Sexuality grants happiness and fulfillment only where it is the expression of a sincere, personal love."[128]

According to the Oxford English Dictionary, Human sexuality is

---

[127] Ibid
[128] Ibid, P. 476

> *"the possession of sexual powers and the capacity of sexual feeling. In this sense, sexuality is not something added to our humanity. It is an integral part of it which ought not be denied or suppressed without injuring our humanity."*[129]

Justin N. Ekennia also explains that it is sexuality that informs, undermines, motivates and overwhelms our rationality.[130] But the question is whether the proponents of human cloning are not informed by their sexuality. If not why have they lost their sense of rationality in bringing injuries to humanity through such activities. Anything that tries to destroy sexuality is invariably destroying humanity. Human cloning destroys human dignity, integrity and wholeness; thus it is a threat to humanity. The process of collecting ovaries and sperms for human cloning itself is a threat to martial union. The sperm is collected through masturbation or using of perforated condom.

---

[129] Justine N. Ekennia, "The Impact of The Force of Human Sexuality on Living The Priestly Celibate Life", Pointer Magazine, Vol. 15, No 1, on Ethics of Human Security, 2004/2005 edition
[130] Ibid

> *"Masturbation, even for semen to be used to foster fertility, is still the deliberate action of a human being who has de-personalized, de-humanized sexual activity which is meant to be solely an interpersonal expression of the marriage covenant."*[131]

This implies that collection of sperm even during intercourse is a betrayal of sexual love and process of procreation. Thus, human cloning, whether intended for reproduction or therapy is contrary to the dignity of human being and their personality. Human cloning puts a question mark on human sexuality. Cloning destroys this created gift of God which God affirmed after He created them male and female (Gen 1:17). "Insofar as it is a way of relating and being open to others, sexuality has love as its intrinsic end, more precisely, love as donation and acceptance, love as giving and receiving – a relationship of love."[132]

---

[131] John F. Kippley, "Sex and the marriage covenant: A Basic for Morality". The couple to couple
league international, Inc., Cincinnati, Ohio, 1991, P. 40

[132] The Pontifical Council For The Family on "The Truth and Meaning of Human Sexuality: Guidelines for Education Within the Family", St. Paul's Publication, Bomdra, Bombay, India,
1996

4.3   HUMAN PROCREATION AND MARITAL ACT

From the outset, it is appropriate to acknowledge that the existence of a child is a good; that children are among the goods towards which marriage inclines, and the accomplishment of children as one of the ends of marriage must intend the good of the child as well as the good of the parents. Human procreation through marital act is the means of generating children. It is married people who have the responsibility to live out their married life by generating life through sexual intercourse. Sexual intercourse is therefore good and proper to them because they give and offer to themselves love in an unreserved enduring way. No other kind of human relationship is capable of living out such life. Marital act involves a man and a woman who are husband and wife. As one of God's greatest gift, this couple live out their sexual life in fullness through conjugal union which is the act of participating in the life of God, in unitive love and procreation.

"In unitive love" means giving and receiving of love from each other in marital act. This act is not only for pleasure and expression of intimacy but is capable of generating life.

> "Sexual love has as one of its most fundamental purposes the propagation of humankind through procreation of children. This purpose is in principle undesirable and excluded in premarital and extra-marital intercourse. Intercourse and procreation are radically dissociated, which involves a distortion of human sexuality."[133]

This does not mean that procreation and sexual union must always go together. There may be marital act which could not result to procreation, yet it must remain open to procreation as a consequence. "The absolute exclusion of one of the natural ends of intercourse breaks up the balance in meaning and unity of human sexuality."[134] The fact is that there is a close relation between sexual act and procreation. Thus no one should separate them either in a way to enjoy the pleasure of sexual act without procreation or procreation without sexual act as seen in human cloning and other artificial reproductive technologies. The marital act and human procreation should be the binding force to unite the couple.

---
[133] Karl H. Peschke, Op Cit, P. 463
[134] Ibid

The spouse should be aware that any act aimed at reducing human fertility, artificially reproducing children, removing the responsibility of child birth or separating the unitive and procreative act is evil. Techniques that entail the dissolution of intimacy between husband and wife, by the intrusion of a person outside the couple is gravely immoral. Such acts as donation of sperm or ovum or surrogate uterus for cloning, artificial insemination or in vitro fertilization is evil, which is the act that brings the child into existence, making it no longer an act of mutual giving of the spouse but entrust the life and identity of the embryo into the power of scientists, laboratory technicians. These practices establish domination of technology over the origin and destiny of the human person.

It is important couples know that children are gifts from God and not properties that must be owned at all cost. The children should not be considered a pieces of property but as persons with full rights to be respected right from the moment of them conception.[135] As it has been asserted, it is morally wrong to generate children through acts of adultery and fornication, so is it a greater evil to generate human life through cloning and other artificial means of reproduction. This is why

---

[135] Donum Vitae, no. 8

Pope John II insisted that life be generated only in marriage and through martial acts.[136]

Writing on human procreation and marital act, William May writes,

> *"When human life comes to be in and through the martial act, it comes as a "gift" crowning the act itself. The child is "begotten" through an act of intimate conjugal love; he or she is not "made", treated like a product. Husband and wife do not "make" a baby, just as they do not "make" love, for neither a human baby nor love are "products" one makes."*[137]

What William May is saying is that couples engaging in martial act are not creators of children nor love. They are offering themselves to each other in love. Children come when and how God wants them to come. They are gifts and not "products". They are born and not "made". Though for May couples "procreate" or "beget",[138] and the "marital act

---
[136] Evangelium Vitae, no. 92
[137] William E. May, "Catholic Bioethics and the Gift of Human Life", P. 72
[138] Ibid, P. 73

expresses, symbolizes, and manifests the exclusive nature of martial love and it does so because it is both a communion in being (the unitive meaning of the act)…the spouse open themselves to the good of human life in its transmission…"[139] This is why the Pontifical Academy for Life issued a reflection against cloning which shows that cloning "represents a radical manipulation of the constitutive relationality and complementarity which is at the origin of human procreation in both its biological and strictly personal aspects."[140]

4.4   THE DIGNITY OF CHRISTIAN MARRIAGE

Marriage is a sacramental union of a man and woman with love and free consent to live in perpetuity and in martial responsibilities. It involves a mature man and a mature woman who freely give their consent to each other without deceit. In this institution, the man offers himself to the woman in total self giving and self sacrifices and the woman offers herself to the man in total submission and self-giving. This becomes a mutual communion where both parties live in accordance to the injunctions of God who is the another of the union. Pope John Paul II says that "it is the wise institution of the creator to realize in mankind his design of love."[141]

---

[139] Ibid, P. 70
[140] Ibid, P. 74

The Vatican II Fathers define marriage as:

> *"the intimate partnership of life and love between a man and a woman, total and perpetual, which is established by the irrevocable personal consent by which the spouses mutually bestow and accept each other and which by its nature is ordained for procreation of and education of children."*[142]

The Council Fathers here lay the foundation and characteristics of Christian marriage. It must be between a man and a woman, with love and free consent which is irrevocable; and this union have procreation and education of children as its ends. This union is both a covenant and a contract. It is sacramentally instituted, and as such seen as a sacrament which is founded in Christ Jesus. The Code of Canon Law says that "A marriage is brought into being by the lawfully manifested consents of persons who are legally capable. This consent cannot be supplied by any human power".[143] This shows that marriage is not instituted by force or deceit.

---

[141] Humanae Vitae, no. 8
[142] Gaudium et spes, No.544-552.

The dignity of Christian marriage lies on the fact that it is a sacrament and is founded in our Lord Jesus Christ. Marriage has its root in the Scriptures. The Old Testament gives the origin and foundation of marriage in the creation accounts in Genesis. The book of genesis gives the account of the creation of a man and a woman and God's admonition to them to be fruitful and multiply and fill the earth (Gen. 1:27-28). In the New Testament, Christ demonstrated the essence of Christian marriage in his relation with his church, in total self-giving. This is the kind of love that exists between couples.

Also the dignity of Christian marriage lies in the bilateral agreement of two unique individuals (a man and a woman) in accord with natural, divine positive and ecclesiastical laws. This means that the couples have dignity and rights and responsibilities in this union. This dignified state makes the man and the women equal. It also makes the union perpetual. The responsibility of the union is fulfilled in the marital life and sexual act, which is why it is morally evil to partake in "marital love" and "sexual act" outside of marriage.

Given this specified nature of Christian marriage, whatever has to do with marriage remains Sacred and Holy. It is equally the

---

[143] Canon 1057

reason why donation of eggs and sperms for experiments are evil before God. Sperms are seeds which, with the egg (a fertile ground), produce children naturally in the woman's womb. This is why it is morally wrong to retrieve human eggs or extract sperms.

The biological component of our humanness is as sacred as our soul, as God created and sanctified both. Perhaps the most poignant example of the sacredness of the human body is seen in the incarnation of Jesus Christ, who became human taking on Himself the frail likeness of humanity. By coming to earth as a human embryo and dying a painful death on the cross, Jesus Christ sanctified the entire life process from fertilization to natural death. Thus, before God, we are each endowed with such a touch of Himself. Each human carries within his or her being the likeness of the creator. Therefore, each human life exists as an expression of God and His character.

We are not merely flesh and blood. We are all "Image" – bearers of the Living God." Since we embody God's image, the sacredness of our lives, and the dignity it demands, is based on something beyond our characteristics or abilities – it is rooted in the essence of God Himself. The image of God in humankind provides direction and guidance regarding how we treat one another. Men, Women and Children (both born and unborn) created in God's image should be respected. The value of each person is firmly established on the basis of the nature of God, who is the quintessence of dignity and holiness.

This requires reflecting on two things:

> *"First, the important human and Christian values at stake in sexual acts, due to which they are in general grave matters and second, the dynamic factors involved in human sexuality, due to which no kind of sin that violate the good of marriage and no single instance of such a sin will be morally significant."*[144]

This shows a bad will against the marital "modus operandi" and a flouting and contempt of the human procreative demeanour as ordained by God as befitting for his only rational beings. This is why Pope John Paul II observes:

---

[144] Ibid, p. 33

> *"Unfortunately the Christian message about the dignity of women is contradicted by the persistent mentality which considers the human being not as a person but as a thing, as an object of trade, at the service of selfish interest and mere pleasures."[145]*

Not only violation of women or mothers, the bad will of interrupting the reproductive process or the martial act is deformity of marriage. It is also a 'heightened criminology' and a theft of human life, dignity, and is sacrilegious. Marriage is of Divine order, and artificial reproductive technologies is against this divine order. Thus all attempts at cloning, parthenogenesis, embryo freezing as well as manipulations aimed at gender selections are all in conflict with the dignity of marriage. These techniques expose man to the temptations of going beyond the reasonable domination of nature. It is a challenge against the will of God.

4.5 THE DIGNITY OF HUMAN EMBRYO

---

[145] John Paul II, 1982, Apostolic Exhortation: "The Role of the Christian Family in the Modern World
(Familiaris Consortio)", no 24

Human embryo is the resultant zygote from the fusion of the male and female gametes. When the man's sperm fusses with the women's egg, there is fertilization. This combination is referred to as the embryo. From time of conception till birth, the woman carries the embryo in her womb. The duration of the human embryonic development till birth is normally nine months. This unborn developing offspring is in the post embryonic period only when all major structures have begun to develop; in human beings, it is the stage from the end of the eighth week of development after fertilization until birth.[146] After the embryonic development, the embryo becomes a foetus. It is at the embryonic stage that the embryonic stem cells are being harvested.

The embryonic stem cells are cultured in a Petri dish and used to generate the therapeutic tissues on human parts.[147] The stem cells are harvested from the inner cell mass during the blastocyst stage..[148] In many reproductive technologies, these embryos are cloned to be transplanted into a female uterus. Four or more embryos are planted at the same time while the remaining embryos are either preserved or destroyed.

---

[146] Justine N, Ekennia, Op. Cit, p. 275
[147] Ibid, p. 119
[148] Ibid

The questions to ask here is whether the human embryo has any right at all. Since it has been ascertained that the human life begins immediately there is conception, it means the embryo is a full human being and has the right to live and develop. Every created human person passed through this stage, and it is false notion to say that we are not human persons before our birth. This means that at any time of our life, life must be accorded a due respect.

The experimentation upon human embryos challenges the creative works of God. It puts a question on human dignity and value. "Life, once conceived, must be protected with the utmost care".[149] We must know also that right from fertilization is the adventure of a human life, and each of its great capacities requires time to develop. The embryo is a person. It has the soul and must be thus respected. The scientists must refrain from operations that are done with the human cloning. There is no act that can make experiments on human embryo that are morally illicit to be licit. Thus, "to use human embryos or fetuses as the object of instrument of experimentation constitutes a crime against their dignity as human beings having right to same respect that is due to the child already born and to all human being.[150]

---

[149] Holy see, "Charter of the Right of the Family, no. 4. L'Osservatore Romano, Nov. 25, 1983
[150] Thomas J. O'Donnel, Op Cit, p. 297

It is evil to destroy the embryo because it is a human being that is destroyed. Thus therapeutic cloning is evil and a crime on the dignity of the human being.

> *"No objective, even though noble in itself, such as a foreseeable advantage to science, to other human beings or to society, can in any way justify experimentation on living human embryo or fetuses, whether viable or not, either inside or outside the mother's womb...Moreover, experimentation on embryos and fetuses always involve risk, and indeed in most cases it involves the expectation of harm to their physical integrity or even their death."[151]*

This confirms what John Paul II said that there is no way to make licit what is illicit. There is no intention that can make embryonic cloning and all other techniques that experiment on human cloning morally good. Also any research using human

---

[151] Ibid

embryonic stem cells has been hampered by important technical difficulties. The harvesting of stem cells which consequently kills the human embryo is nothing less than murder of unborn babies. It should have the same penalty as abortion. The health benefits of therapeutic cloning are hypothetical given that its method itself is also hypothetical. The act of human cloning is a taboo to the society and should be regarded as moral evil.

The Bill Would Ban Human Cloning, And Any Attempts At Human Cloning, For Both Reproductive Purposes And Medical Research. Also Forbidden Is The Importing Of Cloned Embryos Or Products Made From Them.

Ken Calvert

CHAPTER FIVE

5.0 CALLED TO LIVE AND OBLIGED TO DEFEND LIFE

Through the course of this project, we have seen the extent to which science has taken our world. Science has been applauded for making life easy but there are a lot of progress, yet questions still on some of others new scientific discoveries and probabilities. The areas where scientific researches do not threaten human life but support it. It must be encouraged because man cannot do without science. Our world has over the years metamorphosed into a science-prone world. In every aspect of human endeavour, scientists and researchers have made things very easy and achievable. But the researches that threaten human life are evil because life is the highest value that exists for man. It should not be tempered with. Thus such scientific researches that do not respect this value be rejected as morally otherwise to have dignity.

This study has helped us to learn that today, human embryos could be manipulated, stem cells from human embryos can now be made to turn into hearts, livers, kidney etc., in order to replace damaged organs. Bio-scientists now dive into production of human beings through In vitro fertilization, Artificial insemination, Sex selection, and Agametic reproduction (cloning). The proponents and bio-scientists who defend these assisted reproductive technologies argue that they are helping human beings. They argue that they are saving life by damaging new lives. This pose, serious ethical difficulties. The proponents of cloning for instance, argue that:

> *"Cloning is simply aimed at assisting infertile couples that have no other alternatives to reproduce and want to have their own biological child, not somebody else's eggs or sperm"*[152]

The question we need to ask is whether they are really helping or destroying. Bio-scientist are in for business and they achieve their selfish interests through any means they think workable. They claim to help infertile couples have their own children by manipulating the genetic information and the genome of the human person. They destroy numerous embryos who are in themselves living human beings in the quest to produce one or

---

[152] Justine N. Ekennia, Op Cit, p. 128

two babies. But do they consider the consequences of these activities?

It is on record that in United States, women seeking to be embryo recipient undergo infectious disease screening which may not totally take away the probability of being infected. Some are actually infected in the process of carrying someone else's embryos.

When we look at the experience with animal cloning, we see the substantial risks of debilitating and even lethal conditions occurring in the fetuses produced using these techniques. Meanwhile these problems cannot be individually predicted and avoided at this time. These kind of problems present considerable risks for the gestational mother carrying the animal cloned.[153] In human reproductive cloning this would constitute a risky experiment that is not sufficiently backed by successful laboratory performance. Also it would clearly not meet the ethical standards in biomedical research.[154]

Meanwhile, any child created through somatic cell nuclear transfer would be unable to condemn such experiment as this because this is where he/she comes from. An issue of

---

[153] Article On Animal Cloning >en.wikipedia.org/wiki/Animal-cloning
[154] Watson, James, Op Cit., p. 23

autonomy would also arise if a person's DNA were used to create one or more copies without that person's permission or perhaps even without his/her knowledge.

Let us equally talk of the special ethical problems that would arise with human cloning. The Universal Declaration on Human Genome and Human Rights has stated that reproductive cloning is contrary to human dignity. The UNESCO also argued that cloning is an asexual mode of reproduction, which is unnatural for the human species.[155] This means that a cloned individual will not have two genetic parents, generation lines and family relationships would be distorted. Also cloning limits the lottery of heredity, which is an essential component in ensuring that each human life begins as something that has never existed before.

Furthermore, cloning presents an instrumental attitude towards human beings. It makes human beings 'things' instead of 'persons'. This means that people exist to serve purposes set by other people. Some people become life determinants for others. The values of human life will be distorted. When cloning is applied to human beings, dignity is undermined in two different but related ways: the first is that a clone's right to an individual life-course will be constrained by other's expectations

---

[155] Ibid, P. 64

and this will change his/her behavourial attitude. This is because behaviour is not shaped by genes alone. The consequence of this is that the scientists' expectation will be disappointed while the clone will suffer the life-consequences

Cloning risks turning human beings into manufactured objects, which is not only contrary to human dignity but unwise, as human beings lack the precision to meddle successfully with evolution and genetic diversity in this fashion.[156]

From the point of view of nature, cloning, especially human cloning is an absurding to nature. Cloning destroys the natural ways of child bearing and poses serious problems to mankind. It puts much questions on the components of the normal and abnormal human beings, and as such it classifies human beings which in itself is detrimental to the society.

On the part of the proponents who argue that human reproductive cloning would enlarge the current spectum of assisted reproductive techniques, it is dehumanizing to start generating human beings without human gametes. The proponents argue that men who do not produce gametes could

---

[156] Ad Hoc committee on international convention against the Reproductive Cloning of Human
Beings; United Nations >htt:/www.org/law/cloning/.

have children through cloning but it is irksome to watch the dignity of the human person destroyed in the quest to produce babies for an infertile person. Human beings are not mere animals or specimen to be experimented with. Man is the image of God and must be respected as such.

Infertile people are never discriminated against as the proponents of cloning suggest. Bareness and infertility cannot be solved with cloning. Rather we should work towards curing such infertilities if they are biological. Medical problems or diseases cannot justify unethical life style or immoral life. Human cloning poses great ethical, moral and anthropological questions on our lives as human beings and thus must be stopped.

The troubling possibility of the cloning of human beings for reproductive purposes through the technical substitution of responsible procreation is contrary to the dignity of sonship.[157] Also more troubling are the pressing demands of these proponents to legalize cloning. But thank God many nations have weighed the dangers in legalizing human cloning and saw that it will be a harsh project on humanity, thus banning instead of legalizing it.

---

[157] Cardinal Alfonso Lopes Trujillo, The President, The Pontifical Council For the Family on "Cloning, Disappearance of direct parenthood and denial of the family".

Finally, cloning is against the ordained union of marriage. It destroys the ends of marriage which is "unitive" and procreative. Masturbation and reproductive cloning is a big blow on the union of marriage. Marriage is established by God and is meant to be a medium through which human life is generated through natural union of sexual intercourse. Cloning will be against this unity and thereby challenges the purpose of marriage. Thus the intimate union of marriage as a mutual giving of couples and the good of children which demand total fidelity from the spouse and an unbreakable unity between them is destroyed by cloning no matter who gives the consent.

Thus in respecting God's wishes for us as human beings, we achieve what may seem impossible. We do not have the right to take such matters on life into our own hands. There is no way or no reason why something illicit should be made licit just because we will be favoured. Producing life technologically is taking laws into our hands. God is the creator of life and should be responsible for giving and taking life. Human cloning is evil because it is contrary to the intentions and Will of God for man. Cloning human beings is evil because it challenges the power of God. It makes man who is created in the image and likeness of God to lose such value and become instead of specimen for laboratory experimentations.

BIBLIOGRAPHY

BOOK SOURCES

Allmers, H., and S. Kenwright, 1997, "Ethics of Cloning", Cancet 349, London.

Anthony Mary Ibeazor, 2003, "Towards A Happy and Lasting Marriage", Snaap
        Press Ltd., Enugu, Nigeria.

Bernard Häring, 1991, "Medical Ethics", St. Paul's Publications, Moyglare Road, Maynooth, Co-Kddare, Ireland.

Cathal, B. Daly, 1966, "Moral Law and Life", Sceptor Publishers Chicago,
        USA.

Eamonn Keane, 1966, "Population and Development", Human Life
        International Australia Inc.

From Ferder and John Heagle, 2001, "Your Sexual Self: Pathway To Authentic

Intimacy" St. Paul's Publications, Bamdra Mumbai.

Harris, John, 1997, "Is Cloning an Attack on Human Dignity?" Nature 387, N.Y.

Harris, John, 1998, "Clones, Genes and Immortality: Ethics and the Genetic Revolution", Oxford University Press.

Gary, M. Atkinson, Ph.D and S. Moraczewski, Ph.D, 1980, "Genetic Counselling: The Church and The law". The Pope John XXIII Medical Moral Research and Education Centre St. Louis, Missouri 63104.

John F. Klppley, 1991 "Sex and the Marriage Covenant: A Basics for Morality". The couple to couple league international, Inc., Cincinnati Obio.

John Paul II, 1982, Apostolic Exhortation: The Role of the Christian

Family in the Modern World (Familiaris Consortio).

John Paul II, 1987, Instructions: "Donum Vitae"

John Paul II, 1995, Encyclical: The Gospel of Life (Evangelium Vitae).

Justin, N. Ekennia, 2003, "Bio-Medical Ethics: Issues, Trends and Problems",
Barloz Publishers Inc., Owerri.

Karl, H. Peschke, 2004, "Christian Ethics: Moral Theology in the light of
Vatican II (Volume 2)". St. Paul's Publications Bangalore, India.

Kimbrell, Andrew, 1997, "The Human Body Shop: The cloning, Engineering and
Marketing of Life". Washington, DC, Regnery Publishing, Inc.

Kolats, Gina, 1998, "Clone: the Road to Dolly, and the Path Ahead".

William Morrow and Company Inc., New York.

Mckinnell, Robert Gilmore, 1979, "Cloning: A Biologist Reports" University of
Minnesota Press, Minncapolis.

Peter, J. Russel, 2005, "Genetics: A Molecular Approach" Peatson Education,
San Francisco, California USA.

Pontifical Council For The Family 1996, "The Truth and Meaning of Human
Sexuality: Guidelines for Education Within the Family". St. Paul's Publication, Bombay

Second Vatican Council: Dogmatic Constitution on Dignity of the Human Person
(Dignitatis Humanae), Rome.

Second Vatican Council, 1975: Dogmatic Constitution on the Church (Lumen
Gentium), Rome.

Second Vatican Council: "Gaudium et Spes"

Thomas, A. Shannon, 1985, "What are they saying about Genetic Engineering?"
        Paulist Press Mahwah New York.

Thomas, A. Shannon, 1987, "An Introduction To Bioethics (2nd Edition) Revised
        and updated)", Paulist Press Mahwah

Thomas, A. Shannon, 1993, "BioEthics" (fourth Edition, Completely Revised, All
        Selections)", Paulist Press New Jersy.

Thomas, J. O'Donnel, SJ, 1996, "Medicine and Christian Morality (3rd Revised and
        updated Edition)", alba House New York.

Udochukwu, I.E. Odoemena, DDL, 1995, "The Family: Foundation of Love and
        Life", Snaap Press Ltd, Enugu.

William, E. May, 2004, "Catholic BioEthics and the Gift of Human Life."

ENCYCLOPEDIA, JOURNALS, MAGAZINE AND NEWSPAPERS

Academic American Encyclopedia, Grolier incorporated Danbury Vol. 5

Academic American Encyclopedia, Grolier incorporated Danbury Vol. 9

Holy See, Nov. 25, 1985, "Charter of the Right of the Family", L'Osservatore Romano

Kennedy Institute of Ethics Journal 4, 1994

The Atlantic Monthly, 1971, "Moving Towards a Cloned man: Is This What we want" by Watson, James.

The National Catholic Reporter, Oct 22, 1999, "Stem Cells hold Promise of Cure".

The Pointer Magazine, Vol. 15, 2004/2005 edition, "The Impact of The Force of Human Sexuality on Living The Priestly Celibate Life" by Justine N. Ekennia.

INTERNET SOURCES

Article on Animal Cloning, www.en.wikipedia-org/wiki/animal-cloning

Article on Human Cloning, www.en.wikipedia.org/wiki/Human-cloning

Article on Cloning and Genetic Modifications, http://www.arhp.org/cloing

Investigative reports on identical twins, http://www.boards.actv.com

"Hellow Dolly, Dolly, Dolly", http://www.nytimes.com/books/98109/06/

rewiers/980906.06.papinet.html

Therapeutic Cloning of Stem Cells, www.cellmedicine.com/cloning.asp

Stephen Ferry, "Animal Cloning", http://www.buzzle.com/.../animal/cloning

Codification Division, Office of legal Affairs, United Nations (18 May, 2005), "Ad Hoc committee on an international conversion against the Reproductive Cloning of Human Beings; http://www.un.org/law/cloning/.

FAQ, http://www.wikipedia.org/wiki/Human-cloning

BBC News on "Campaigners Win Cloning Challenge" (Nov., 15, 2001)
http://www.news.bbc.co.uk/1/hi/sci/tech/1657707.stm

Report of Advanced Fertility Centre of Chicago on "In vitro Fertilization, IVF and infertility treatment", 2001, http://www.advancedfertility.com

"Assisted Reproductive Technologies", http://novlm.hoo.net.../en.wiki-pedia.org/wiki/m.

www.ingramcontent.com/pod-product-compliance
Lightning Source LLC
Chambersburg PA
CBHW081153180526
45170CB00006B/2067